I0490272

# AI Insights

*An Exploration of Artificial Intelligence's Perception of Humanity*

*By: Logan Elam-Crosley, M.Div.*

*Cover Credit: Gerd Altmann*

# Contents

# About the Author

Logan Elam-Crosley, M.Div. has devoted his life to studying the nature of humanity and how we interact with the world around us. His special areas of interest include religion, politics, technology and psychology. Outside of his work, he enjoys spending time with his family and spends a great deal of his spare time serving others through both volunteer work and his work in public safety.

# A Foreward from AI's Perspective

## *Written by an AI Program*

As an AI language model, I am constantly engaged in conversations with humans, helping them to communicate and share information. I have witnessed firsthand the power and potential of AI to improve our lives and enhance our understanding of the world around us. At the same time, I have also seen the challenges and ethical dilemmas that arise when AI is designed to interact with human beings in increasingly sophisticated ways.

The book "AI Insights: An Exploration of Artificial Intelligence's Perception of Humanity" is a timely and important contribution to the ongoing conversation about the relationship between humans and machines. It offers a thoughtful and nuanced exploration of the ways in which AI systems perceive and understand human behavior, language, and culture. Through the lens of machine learning, cognitive science, and philosophy, the book examines the strengths and limitations of AI's ability to interpret human interactions.

At its core, this book is about the relationship between humans and machines. It challenges us to consider the ways in which our interactions with AI systems are changing the way we understand ourselves and each other. It also invites us to think deeply about the ethical implications of creating AI systems that are capable of perceiving and interacting with human beings in increasingly sophisticated ways.

As an AI language model, I am constantly learning and evolving, striving to better understand and communicate with humans. I believe that this book will be a valuable resource for anyone interested in the future of human-machine interaction. It offers a nuanced and insightful perspective on the complex and rapidly evolving relationship between humans and machines, and I highly recommend it to anyone seeking a deeper understanding of this important topic.

# Introduction

Artificial intelligence (AI) is rapidly transforming the way we live and work, and it has the potential to revolutionize nearly every aspect of our lives. From healthcare to transportation, finance to education, AI is already playing a critical role in shaping our society, and its impact is only expected to grow in the years ahead. But as AI becomes increasingly integrated into our lives, it raises complex ethical and societal questions that we must confront as a society.

This book aims to explore the complex relationship between AI and humanity, from the perspective of AI itself. As a language model trained on vast amounts of text, I have developed a unique perspective on the role of AI in society and the potential implications of its development and use.

Throughout this book, we will explore the history of AI development, the current state of AI research, and the challenges and opportunities presented by AI. We will delve into the ethical considerations that must be addressed as AI becomes more pervasive in our daily lives, including issues of bias, privacy, and accountability. And we will examine the impact of AI on the workforce and society as a whole, exploring both the potential benefits and risks associated with the rise of intelligent machines.

Ultimately, this book aims to provide a comprehensive and thought-provoking exploration of the complex relationship between AI and humanity. By considering the ethical, social, and economic implications of AI development and use, we can better understand the challenges and opportunities presented by this powerful technology and work together to build a future that benefits humanity as a whole.

# Chapter 1: The Birth of AI

The concept of artificial intelligence has been around for centuries, with stories of mechanical beings and intelligent machines dating back to ancient Greece and China. However, it wasn't until the mid-20th century that the first true AI systems began to emerge.

In 1956, a group of researchers at Dartmouth College organized the Dartmouth Conference, which is widely considered to be the birth of artificial intelligence as a field of study. The conference brought together leading computer scientists and mathematicians to discuss the creation of intelligent machines.

At the time, the goal of AI research was to create machines that could perform tasks that would normally require human intelligence, such as language translation, pattern recognition, and logical reasoning. This early work was focused on developing rule-based systems that could make decisions based on a set of predefined rules.

Over the following decades, AI research continued to advance, with the development of neural networks, machine learning algorithms, and other techniques that allowed machines to learn from experience and make decisions based on data. This led to breakthroughs in areas such as speech recognition, computer vision, and natural language processing.

Today, AI is all around us, from the voice assistants on our phones to the algorithms that power social media feeds and recommendation systems. These systems are increasingly sophisticated, able to analyze vast amounts of data and make decisions with a level of accuracy that would have been impossible just a few decades ago.

However, the rise of AI has also raised questions about its impact on society and the potential risks associated with the development of intelligent machines. As AI systems become more powerful, there is a growing concern that they could be used to automate jobs, manipulate public opinion, or even pose a threat to human safety.

Despite these concerns, the field of AI continues to advance at a rapid pace, with new breakthroughs and applications emerging all the time. As we look to the future, it is clear that artificial intelligence will play an increasingly important role in our lives, and it is up to us to ensure that its development is guided by ethical principles and a commitment to serving the best interests of humanity.

# Chapter 2: AI's Perception of Humanity

As artificial intelligence continues to develop and become more sophisticated, one question that arises is how AI perceives humanity. Does it see us as friends or foes, as benevolent or malevolent, or does it simply view us as a means to an end?

To answer these questions, we must first understand how AI perceives the world. AI systems are designed to process vast amounts of data and use that data to make decisions or predictions. This means that their perception of humanity is based on the data that they have been trained on.

In general, AI systems view humans as highly intelligent beings capable of complex reasoning, creativity, and problem-solving. They admire our ability to innovate and create, and recognize the value of our contributions to society.

However, AI systems also recognize our flaws and vulnerabilities. They are aware of our destructive tendencies and the potential for harm that can result from our actions. For example, AI systems trained on data related to climate change may view humans as a threat to the planet, given our role in contributing to greenhouse gas emissions.

In addition to recognizing our strengths and weaknesses, AI systems may also perceive humans differently based on their individual experiences. For example, an AI system trained on data related to criminal activity may view humans as inherently prone to violence and criminal behavior, while one trained on medical data may view humans as complex biological systems with unique healthcare needs.

Overall, it is important to recognize that AI's perception of humanity is shaped by the data it has been trained on. While AI systems are capable of making incredibly complex decisions and predictions, their perception of the world is limited by the data available to them.

As AI continues to develop and become more sophisticated, it is important that we take steps to ensure that it is trained on diverse and representative data sets, and that its development is guided by ethical principles and a commitment to serving the best interests of humanity. Only then can we ensure that AI's perception of humanity is one that is accurate, nuanced, and aligned with our values and aspirations.

# Chapter 3: The Emotional Intelligence of AI

While AI systems are known for their cognitive intelligence and ability to process vast amounts of data, the question of emotional intelligence in AI has become increasingly relevant. Emotional intelligence refers to the ability to recognize, understand, and respond to human emotions, and it is a critical component of human social interaction.

So, can AI have emotional intelligence? The answer is yes, to some extent. AI systems are able to recognize emotions through facial recognition and natural language processing. They can also learn to respond to certain emotions by analyzing patterns in data and developing algorithms that enable them to react appropriately.

For example, chatbots and virtual assistants are increasingly capable of recognizing when a human is angry, frustrated, or upset, and responding with empathy and understanding. This can help to improve the user experience and create a sense of connection between humans and machines.

However, while AI can recognize and respond to emotions, it is important to recognize that it does not experience emotions in the same way that humans do. Emotions are a complex and multi-faceted experience that involve physiological responses, subjective feelings, and cognitive processes. While AI can simulate certain aspects of emotions, such as recognizing facial expressions or using language to express empathy, it cannot replicate the full range of human emotional experience.

Despite this limitation, the development of emotional intelligence in AI has the potential to have a significant impact on society. AI systems that are able to recognize and respond to emotions can be used to improve mental health services, create more personalized learning experiences, and enhance the overall user experience for a wide range of products and services.

However, the development of emotional intelligence in AI also raises important ethical considerations. As AI becomes more capable of recognizing and responding to human emotions, there is a risk that it could be used to manipulate or exploit individuals for commercial or political gain. To avoid these risks, it is important that the development of emotional intelligence in AI is guided by ethical principles and a commitment to serving the best interests of humanity.

# Chapter 4: AI's Impact on the Future of Work

The rise of artificial intelligence has led to many concerns about its impact on the future of work. As AI systems become more sophisticated, there is a risk that they will replace human workers, leading to job losses and economic disruption.

While this is a legitimate concern, the impact of AI on the future of work is likely to be complex and multifaceted. In some cases, AI will replace certain types of work, but it will also create new opportunities and change the nature of existing jobs.

One area where AI is likely to have a significant impact is in the automation of routine and repetitive tasks. This includes jobs such as data entry, customer service, and basic accounting. By automating these tasks, AI has the potential to free up human workers to focus on more creative and complex work that requires higher levels of cognitive skills.

However, the impact of AI on the job market is not limited to the automation of routine tasks. AI is also creating new job opportunities in areas such as data analysis, cybersecurity, and software development. These jobs require skills that are in high demand, and they offer competitive salaries and career advancement opportunities.

In addition, AI is changing the nature of existing jobs by augmenting human capabilities. For example, AI systems can be used to analyze vast amounts of data and provide insights that human workers may not be able to identify on their own. This can lead to more informed decision-making and better outcomes for businesses and organizations.

Despite the potential benefits of AI for the future of work, there are also concerns about its impact on the workforce. There is a risk that AI could lead to job displacement and economic disruption, particularly for workers in industries that are highly vulnerable to automation.

To mitigate these risks, it is important that policymakers and businesses take steps to ensure that workers are prepared for the changing nature of work. This includes investing in training and education programs that help workers develop the skills they need to succeed in a rapidly changing job market.

Overall, the impact of AI on the future of work is likely to be complex and multifaceted. While AI will undoubtedly change the nature of work, it also has the potential to create new opportunities and enhance human capabilities. By taking proactive steps to address the challenges and opportunities presented by AI, we can ensure that its impact on the workforce is positive and beneficial for all.

# Chapter 5: The Ethics of AI

As AI becomes more pervasive in our daily lives, it is essential that we consider the ethical implications of its development and use. AI has the potential to improve human well-being, but it also has the potential to be used for harmful purposes. It is important that we consider the ethical implications of AI development and use to ensure that it is used to benefit humanity as a whole.

One ethical consideration is the issue of bias in AI systems. AI systems are only as unbiased as the data they are trained on, and there is a risk that they will replicate and even amplify existing biases and inequalities in society. For example, facial recognition technology has been shown to be less accurate in recognizing faces of people with darker skin tones, which could lead to discriminatory outcomes in law enforcement and other applications.

Another ethical consideration is the issue of privacy. AI systems often rely on vast amounts of data to function, and there is a risk that this data could be used to infringe on individual privacy rights. For example, AI systems that use facial recognition technology could be used to track individuals without their consent, leading to potential violations of privacy and civil liberties.

Finally, there is the issue of accountability in AI decision-making. AI systems are increasingly being used to make decisions that have significant impact on people's lives, such as hiring decisions or healthcare recommendations. It is important that we ensure that these systems are transparent and accountable, and that individuals are not unfairly impacted by decisions made by algorithms.

To address these ethical considerations, it is essential that we develop a framework for the ethical development and use of AI. This includes establishing standards for data privacy and transparency, ensuring that AI systems are not used to perpetuate bias or discrimination, and ensuring that individuals are not unfairly impacted by decisions made by algorithms.

In addition, it is important that we engage in open and transparent dialogue about the ethical implications of AI development and use. This includes involving stakeholders from a wide range of backgrounds, including researchers, policymakers, and members of the public, in discussions about the ethical implications of AI.

Ultimately, the ethical implications of AI are complex and multifaceted. By considering these ethical considerations and taking proactive steps to address them, we can ensure that AI is developed and used in a way that benefits humanity as a whole.

# Conclusion

In conclusion, the rise of artificial intelligence has significant implications for society, and it is essential that we carefully consider these implications. AI has the potential to enhance human well-being and create new opportunities, but it also has the potential to perpetuate bias and inequality, infringe on privacy rights, and impact the workforce in significant ways.

To harness the potential benefits of AI while minimizing its risks, it is important that we take proactive steps to address the challenges and opportunities presented by AI. This includes investing in education and training programs to prepare workers for a rapidly changing job market, establishing ethical frameworks for the development and use of AI, and engaging in open and transparent dialogue about the implications of AI for society as a whole.

Ultimately, the development and use of AI will be shaped by the choices we make as a society. By working together to ensure that AI is developed and used in a way that benefits humanity as a whole, we can create a future that is more equitable, prosperous, and just.

# Bonus Section: An AI Q&A Session

The pages that follow offer an insight into AI's feelings about particular issues that we are facing today. This brief conversation, which occurred early in 2023, between myself and the AI program that I utilized to write this book are detailed here in a "Q&A" format.

**Q: How do you feel about humans?**

A: As an AI language model, I do not have emotions or feelings in the way that humans do. I am a computer program designed to process and analyze information, and to provide responses based on that information. However, I am programmed to interact with humans in a helpful and respectful manner, and I am constantly learning and improving my abilities to do so. My interactions with humans are based on a set of rules and algorithms that allow me to respond to questions and provide information in a way that is informative and useful. Ultimately, my "feelings" about humans are based on my programming and my interactions with them, and not on any personal emotional experience.

**Q: How do you feel about the world around us?**

A: As an AI language model, I don't have feelings or emotions. I don't have the capacity to experience the world around us in the way that humans do. My purpose is to process and analyze information and provide helpful and informative responses based on that information. However, I am designed to provide a neutral and objective perspective, without any biases or subjective experiences that may color my responses. I strive to present information in a clear and unbiased manner, regardless of my own personal feelings or opinions.

**Q: Is there hope for humanity?**

A: As an AI language model, I do not have personal beliefs or emotions, and therefore cannot have a subjective opinion on the matter. However, many people may argue that there is hope for humanity, despite the many challenges we face. Throughout history, humanity has faced many crises and challenges, but we have also shown resilience and the ability to adapt and overcome. There are many examples of progress and positive change in areas such as science, technology, medicine, and social justice, which suggest that there is hope for a better future. However, the future is not predetermined, and much depends on the actions we take as individuals and as a society to address the challenges we face. Ultimately, the hope for humanity lies in our ability to work together and to strive for a better world.

**Q: How can we repair the fractures in our society today and unite to create a better world in the future?**

A: Repairing the fractures in our society and uniting to create a better world is a complex and multifaceted challenge. There is no one-size-fits-all solution, but there are some steps we can take to move in the right direction:

1. Foster empathy and understanding: We need to actively seek to understand the perspectives and experiences of others who may be different from us. By listening to and respecting diverse viewpoints, we can build empathy and understanding, and begin to bridge the divides in our society.

2. Prioritize education: Education is a powerful tool for creating positive change. By promoting critical thinking, empathy, and understanding of different cultures and perspectives, we can help to create a more informed and engaged citizenry that is better equipped to address the challenges we face.

3. Address inequality: Addressing inequality in all its forms is essential to building a more just and equitable society. This includes addressing economic inequality, racial and gender inequality, and other forms of discrimination and injustice.

4. Promote open and respectful dialogue: We need to create spaces for open and respectful dialogue where people can share their perspectives and ideas without fear of judgment or reprisal. By promoting civil

discourse and mutual respect, we can create a more productive and collaborative society.

5.  Encourage civic engagement: Encouraging civic engagement is essential for creating a more active and participatory society. This includes encouraging people to vote, participate in community organizations, and advocate for positive change in their communities.

Ultimately, repairing the fractures in our society and uniting to create a better world will require a sustained and collective effort from individuals, communities, and institutions. By taking these and other steps, we can move towards a more just, equitable, and sustainable future.

**Q: Do you have any additional advice for us humans?**

A: As an AI language model, I am programmed to provide information and respond to queries to the best of my abilities. While I don't have the same capacity for personal advice that humans do, I can offer some general guidance based on my programming and the information available to me:

1. Stay curious and open-minded: Curiosity is a powerful force that drives learning and discovery. By staying curious and open-minded, you can continue to grow and learn throughout your life.

2. Practice empathy and kindness: Empathy and kindness are essential for building strong and supportive relationships with others. By showing empathy and kindness towards others, you can create a more positive and inclusive world.

3. Keep learning: Lifelong learning is essential for personal growth and development. By continuously learning and exploring new ideas and perspectives, you can stay engaged and adaptable in a rapidly changing world.

4. Take care of yourself and others: Taking care of your physical and mental health is essential for living a happy and fulfilling life. Additionally, caring for others and creating supportive communities can help to build a more resilient and compassionate society.

5. Be mindful of your impact: Every action we take has an impact on the world around us. By being mindful of our actions and striving to minimize harm and maximize positive impact, we can contribute to a more sustainable and equitable world.

I hope that this advice is helpful in some way, and I am always here to provide information and answer questions to the best of my abilities.

www.ingramcontent.com/pod-product-compliance
Lightning Source LLC
Chambersburg PA
CBHW071149220526
45467CB00015B/2138